Hydrogen Safety

Topical Conference at the 2021 AIChE Spring Meeting and 17th Global Congress on Process Safety

Online
18 – 23 April 2021

ISBN: 978-1-7138-5328-2

Printed from e-media with permission by:

Curran Associates, Inc.
57 Morehouse Lane
Red Hook, NY 12571

Some format issues inherent in the e-media version may also appear in this print version.

Printed with permission by Curran Associates, Inc. (2022)

For permission requests, please contact AIChE
at the address below.

AIChE
120 Wall Street, FL 23
New York, NY 10005-4020

Phone: (800) 242-4363
Fax: (203) 775-5177

www.aiche.org

Additional copies of this publication are available from:

Curran Associates, Inc.
57 Morehouse Lane
Red Hook, NY 12571 USA
Phone: 845-758-0400
Fax: 845-758-2633
Email: curran@proceedings.com
Web: www.proceedings.com

TABLE OF CONTENTS

617965 Hydrogen: Three Process Safety Risks Not to Take Lightly

Abstract:

Hydrogen is often stored or processed in facilities that contain other hazardous materials such as flammable liquids, reactive chemicals, or combustible dust. This paper will cover three case studies where the risks presented by the hazardous material were compounded (presenting additional and often unexpected risks) when hydrogen was present. This presentation will provide:

1. Addressing Control of Ignition Sources (Electrical Classification and Hot Work) challenges when flammable vapors (heavier than air) and Hydrogen (lighter than air) are present
2. Ignition /explosive concerns presented by combustible dust in a hybrid environment that includes Hydrogen
3. Reactive Chemistry involving waste streams that generate hydrogen off-gasses

The presentation will focus on techniques for identifying the hazards introduced by hydrogen, establishing an appropriate Basis of Safety for similar operations, and providing references to applicable Recognized and Generally Accepted Good Engineering Practices (RAGAGEP) that provide guidance for these unique Process Safety situations. Case studies that have been investigated by the Chemical Safety Board or received enforcement by US-OSHA will be shared during the presentation.

Unusual Hydrogen Explosions Due to Unanticipated Metal-Water Reactions

Robert Zalosh
Firexplo LLC
Wellesley, MA 02481
bzalosh@firexplo.com

Prepared for Presentation at
American Institute of Chemical Engineers
2020 Spring Meeting and 17th Global Congress on Process Safety
Virtual
April 18 - 22, 2021

Unusual Hydrogen Explosions Due to Unanticipated Metal-Water Reactions

Robert Zalosh
Firexplo LLC
Wellesley, MA 02481
bzalosh@firexplo.com

Keywords: hydrogen explosions, electrochemical reactions, metal-water reactions.

Abstract

Inadvertent generation of hydrogen from ambient temperature metal-water chemical and electrochemical reactions have been responsible for a surprising number of hydrogen explosions. This paper describes three sets of incidents involving different metals in applications that are normally considered free of explosion hazards. The first set of incidents involves wet alkaline incinerator ash with an aluminum content of at least two percent. The second set of incidents involves copper conductor electrochemical reactions in subsea electrical cable. The third set involves electrochemical corrosion reactions in galvanized dry sprinkler pipe. In all three applications, test data are discussed to elucidate the specific reactions and potential preventive and mitigation measures.

1 Introduction

Inadvertent generation of hydrogen from reactive powdered metal-water reactions and molten metal-water reactions is a well-known hazard in metal production and processing facilities. Prevention and protection measures for this hazard are described, among other places, in the NFPA 484 Standard on Combustible Metals [1]. However, inadvertent hydrogen generation from ambient temperature metal-water chemical and electrochemical reactions can also occur in many unanticipated settings and scenarios, and such scenarios have been responsible for a surprising number of hydrogen explosions.

Each of the following three sections of this paper provides details of one incident and then summarizes several other incidents for a given metal and application. It then describes the specific reactions likely responsible for the inadvertent hydrogen generation, including pertinent test data. The first set of incidents involve hydrogen generated from reactions of water with Municipal Solid Waste incinerator ash containing aluminum. The second type of metal-water electrochemical reaction involves copper cable in a seawater atmosphere. The third set of incidents entails hydrogen generation from galvanized steel corrosion reactions in dry pipe sprinkler systems.

2 Water-Incinerator Ash Reaction Incidents

2.1 Nortrader Cargo Vessel Explosion

A prime example of this type of incident is the January 2017 explosion on the dry cargo ship *Nortrader* that was the subject of a comprehensive investigation and report by the British Marine Accident Investigation Branch (MAIB) [2]. The *Nortrader* cargo of unprocessed incinerator bottom ash was generated in a nearby Plymouth, England power plant burning Municipal Solid Waste (MSW). Prior to the January 2017 incident, there had been 34 uneventful shipments of MSW bottom ash on 26 vessels similar in size to the 88 m length, nominal 1900 metric ton capacity *Nortrader*.

On January 12, 2017, over a period of about eight hours, 2,333 metric tons of wet MSW bottom ash were loaded into the *Nortrader's* 3409 m^3 cargo hold during steady heavy rain. After loading, the vessel left the dock and anchored overnight in Plymouth Sound while it waited for the weather to improve. During this time, the cargo hold ventilation fans and inlets were closed because of the heavy rain.

A large propagating explosion occurred the next day while the *Nortrader* was still at anchor. The ignition source for the explosion was the actuation of a stop button on an electrical panel situated in a ship compartment immediately forward of the large cargo hold. The ship's chief engineer, who was standing in front of the electrical panel suffered first degree burns to his face and second degree burns to his body, both hands, and lower extremities. The explosion then propagated into the cargo hold and lifted and bent all nine of the cargo hold steel latch covers, which had been held in place by latch cleats. The displaced hatch covers over the cargo hold the are visible in Figure 1.

Figure 1. *Nortrader* damaged cargo hold latch covers after explosions.

After the incident, the MSW from the Nortrader cargo hold was unloaded for inspection and testing. Employees of a consulting company retained by one of the ship's insurers drilled holes into the unloaded pile of MSW bottom ash and tested for the presence of flammable gas. They detected 1.4% hydrogen and 0.2 % carbon monoxide in the pile void spaces.

The MAIB hired another consulting company to acquire samples of the unloaded bottom ash for two types of water reactivity tests. One test series was UN Test N.5, the test method for substances which in contact with water emit flammable gases. It is described in section 33.4.1.4 (Annex E) of the UN Manual of Tests and Criteria [3]. The second test series involved an ad-hoc test with a much higher water-to-ash ratio (6:1) than in UN Test N.5 (1.5:1). Sample preparation for both tests entailed grinding the sample particulate to less than 250 micrometers for UN Test N.5, and to less than 1 mm for the ad-hoc reactivity test.

Several of the ash subsample water reactivity tests resulted in measured hydrogen production rates between 0.41 l/kg-hr and an average of 0.77 l/kg-hr. These rates are slightly less than the UN Test N.5 threshold of greater than 1 l/kg-hr of flammable gas to classify a material as a Division 4.3, substance which emits flammable gas when in contact with water [3]. Nevertheless, these hydrogen production rates indicate a potential for hydrogen accumulation in cargo holds and other enclosures with negligible ventilation and large quantities of bottom ash. Perhaps the UN Manual test criteria should be re-evaluated to provide a separate, less restrictive, category for this type of material that generates lower rates of flammable gas when in contact with large quantities of water.

The unventilated *Nortrader* cargo hold containing water-soaked bottom ash allowed the slowly generated hydrogen to accumulate over a period of about 19 hours while it was anchored in Plymouth Sound. An unsealed cargo hold light fixture access cover provided a path from the cargo hold to the adjacent ship compartment, which was also unventilated during this period. According to the MAIB report, the electrical panel in this compartment had an unsecured door, which allowed hydrogen to enter the panel interior. When an electrical circuit within the panel was energized and an arc or spark developed across a pair of contacts, the hydrogen-air mixture within the panel was ignited. The resulting explosion propagated from the panel into the ship compartment and then, via open light fixture cover, into the large cargo hold. An analysis by a MAIB contractor concluded that the hydrogen concentration in the cargo hold at the time of the explosion was in the range 6% to 9%, corresponding to a hydrogen volume between 114 m^3 and 171 m^3. The two explosions felt and heard by the *Nortrader* crew were the explosions in the electrical panel compartment and in the cargo hold itself.

2.2 MSW incinerator facility explosions

There have been several reported hydrogen explosions at MSW incinerator facilities. One such incident in Japan in 1995 is well documented [4,5,6]. A large amount of aluminum in the waste stream (9.6% Al) produced a clinker, i.e., a hard chunk of solidified molten metal and bottom ash, that was blocking flow in an ash discharge chute leading from the bottom of the incinerator to an ash hopper. Employees attempted to remove the clinker by striking it with a chisel and intermittently pouring water on it. This caused an explosion with ejected hot gas and ash burning three of the workers, one of whom subsequently died from the burn injuries [4,6]. The explosion was attributed to hydrogen formed from the aluminum-water reaction, with the subsequent hydrogen-air mixture in the chute being ignited by the chisel impacting the still hot clinker and ash.

A similar explosion occurred in a Japanese MSW incinerator in 1983 [6]. This explosion occurred in a dust bunker when an iron bar was used to crush some blocks of ash. The worker suffered severe injuries when he was thrown from the explosion site. The ash moisture content was about 10%, and there were significant amounts of aluminum in the ash.

Larson [7] briefly described a 2003 explosion in the basement ash storage bunker at a Swedish MSW incinerator due to hydrogen generation. Fortunately, there were no injuries, but the explosion damaged the bunker walls. Other incinerator facility explosions in Sweden have occurred during conveyor transport of warm wet bottom ash [8].

2.3 Other ash-water reaction explosions

Arm and Lindeburg [8] mention the occurrence of hydrogen explosions due to ash-water reactions in some Swedish ash storage facilities such as rock caverns and former oil wells. They also mention some explosion incidents associated with wetted fly ash in landfills.

There have been some flash fire and explosion incidents involving hydrogen generated from the ash aggregate used to make lightweight/foamed concrete. One such incident in 2009 in the UK occurred in a former dry well that had recently been filled with foamed concrete [9]. A grinder being used to cut reinforcing rods above the concrete ignited the hydrogen-air mixture that had accumulated there. This and other UK incidents involving foamed concrete and bricks caused the UK Highways Agency to ban the use of foamed concrete made with incinerator bottom ash.

2.4 Ash-water reactions producing hydrogen.

Although a variety of metals can react with water to produce hydrogen, most of the incinerator ash generated hydrogen has been found to be due to aluminum-water reactions. The primary reaction responsible for hydrogen generation is believed to be [10]:

$$2Al + 6H_2O \rightarrow 2Al(OH)_3 + 3H_2 \quad (1)$$

This reaction is promoted in an alkaline environment, with the major alkali source in MSW incinerator ash being calcium hydroxide. According to Nithiya et [10], the calcium hydroxide is consumed and regenerated in the following reactions, which combine to produce reaction (1).

$$2Al + 6H_2O + 2Ca(OH)_2 \rightarrow 2CaAl(OH)_5 + 3H_2 \quad (2)$$

$$2CaAl(OH)_5 \rightarrow 2Ca(OH)_2 + 2Al(OH)_3 \quad (3)$$

The presence of these compounds in ash-water reaction products has been confirmed by Nithiya et al [10], and the promoting effect of pH values > 10 in ash has been demonstrated in test data reviewed by Larsen [7] and others. Several researchers have reported significantly greater hydrogen generation rates when the ash is mixed with alkaline solutions compared to mixing with distilled water. Nithiya et al [10] cite research showing that the alkaline ash conditions cause the oxide layer on the aluminum surface to be perforated and dissolved, allowing direct contact and reaction of aluminum and water.

As implied by reactions (1) and (2), hydrogen generation rates from ash-water reactions increase with increasing aluminum content and increasing pH of the ash. The reactions have also been shown to be faster for freshly generated warm ash, and to decay after several days of ash exposure to ambient air (allowing greater oxide layer formation). These effects produce a wide variation in reported hydrogen generation yields obtained in tests with MSW ash samples.

Besides hydrogen generation rate data, reported test data include total hydrogen generated per unit mass of ash. The hydrogen yield data show the yield increasing with increasing ash aluminum content as well as other factors mentioned above. Yield data have been compiled from a variety of sources including data for bottom ash from Japanese incinerators [10], fly ash from Japanese incinerators [6], and both bottom ash and fly ash from Swedish incinerators [8]. The hydrogen generation data cumulative distribution for ambient temperature reactions obtained by mixing with distilled water are shown in Figure 2. The median hydrogen yield is 2.4 liter-H_2 per kg-ash for bottom ash and 10 l/kg-ash for fly ash. The highest measured hydrogen yields are 24 l/kg-ash for bottom ash and 37 l/kg-ash for fly ash.

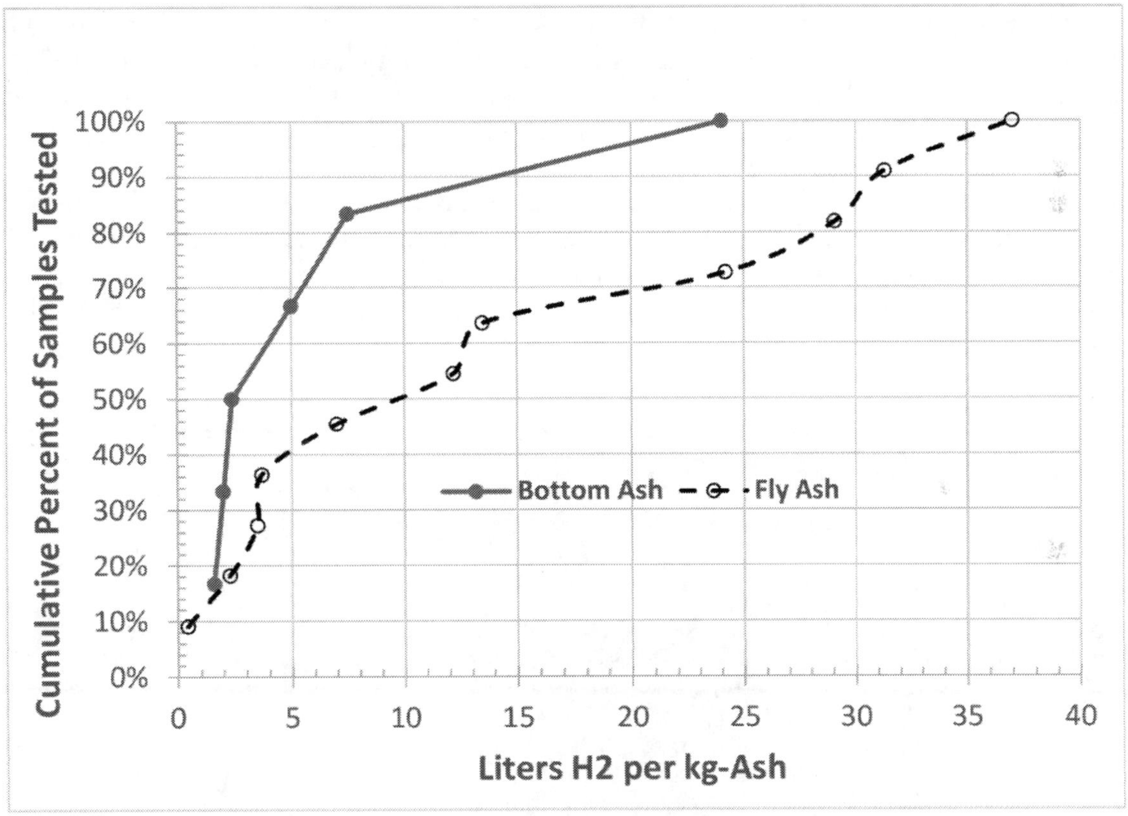

Figure 2. Hydrogen generation data cumulative distribution for bottom ash and fly ash reactions with water.

The data shown in Figure 2 can potentially be used to provide a risk-based estimate of the hydrogen accumulation hazard potential in MSW incinerator ash bunkers and storage and transportation facilities. Daily generation data reported in the cited references show that most of the hydrogen is generated in the first few days after ash formation and contact with copious quantities of water. (Based on the test methods cited above, copious water implies having a water/solid ratio of about two or greater at some point during handling and storage.) Therefore, this data can also be used as a basis for estimating MSW ash storage facility ventilation requirements to prevent flammable mixture formation. These requirements are not needed for ash obtained from other fuels providing those fuels do not have aluminum contents of about 2 percent or higher.

3 Electric Cable – Seawater Reaction Hydrogen Incidents

3.1 Offshore Junction Box Explosions

Offshore oil and gas exploration and production platforms have many electrical junction boxes exposed to sea air and aerosols that produce a saline environment within the box. Several of these boxes have experienced explosions that damage the box. Incident investigations have determined that hydrogen generation within the box produced a flammable atmosphere that was probably ignited by an arc, spark, or electrostatic discharge.

One junction box explosion incident [11] occurred in a box installed above the boat landing level of an unmanned platform in a tropical environment. The box had an IEC 60079 [12] designation "Ex e", meaning that it is intended for normally non-sparking electrical equipment with a rated voltage no greater than 11 kV, to be installed in a potentially flammable gas atmosphere. This particular box contained a 6.6 kV submarine power cable, and had breather ventilation.

During a maintenance crew visit about a year after the cable was first energized, the junction box hinged door was observed to be bulged outward and blown open, as can be seen in Figure 3. There was no interior damage or char deposits and the electrical connections remained intact. The fact that the box did not withstand the explosion is not surprising since it was not constructed or rated as an "Ex d" box.

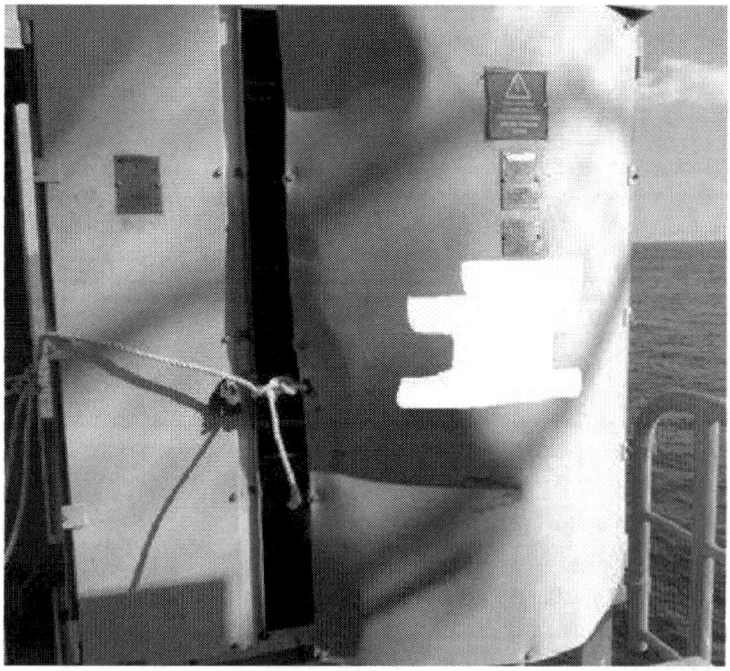

Figure 3. Damaged junction box door on offshore platform [11].

There have been at least two similar reported junction box explosion incidents at other platforms and locations. After one of those incidents, flammable gas was detected in the box [11].

Hydrogen generation from the copper conductor within seawater cables is a recognized phenomenon in subsea electrical cables with the generation and associated copper loss being more pronounced when there is cable insulation damage. This is illustrated in Figure 4, which shows the hydrogen bubbles originating from damaged cables in simulated seawater [13]. Hydrogen is also formed from seawater reactions with the copper and steel components of optical fiber cable [14]. Some seawater cable manufacturers provide hydrogen absorbing materials in their cables to prevent the hydrogen from being released from the cables.

Figure 4. Hydrogen bubbles from a damaged cable in simulated seawater [13]

3.2 Buried Cable Seawater Reaction Hydrogen Incident

An unusual electric cable-seawater reaction hydrogen explosion occurred at a Rhode Island beach in July 2015 [15]. The hydrogen was generated by the seawater corrosion of an abandoned electrical cable that had been used until 2007 to power Coast Guard navigational lights on rock jetties near the beach. People at the beach were not aware of the cable which was buried in the sand. When a hydrogen-air mixture was somehow ignited, the explosion blew a woman out of her beach chair. She suffered broken ribs and bruises upon landing on a jetty about 10 feet away.

Figure 5 shows a portion of the cable after the sand that had obscured it was removed. University of Rhode Island geochemists tested sand samples from the area and measured hydrogen concentrations as high as 5%, i.e., slightly above the Lower Flammable Limit. They also observed that the corroded cable did not have a green patina, indicating that the corrosion was due to oxide, rather than chloride formation.

Figure 5. Corroded electric cable under beach sand [15]

3.3 Seawater Induced Copper Hydrogen Evolution Reactions

Copper exposure to a seawater alkaline environment promotes hydrogen evolution via two different electrochemical reactions. First, copper serves as a catalyst for the cathodic reaction

$$2H_2O + 2e^- \rightarrow H_2 + 2[OH]^- \quad (4)$$

Sherif and Park [17] maintain that this reaction is more likely to occur in de-aerated solutions, whereas aerated NaCl solutions are more likely to produce the water and oxygen reduction reaction

$$O_2 + 2H_2O + 4e^- \rightarrow 4[OH]^- \quad (5)$$

However, Martins et al, [18] have shown that reaction (4) can produce copious amounts of hydrogen when the polycrystalline copper surface allows the formation of a copper-hydroxide activated complex. They demonstrate that copper surface treatment via polishing and annealing allows this activated complex to form and render the copper surface more hydrophilic.

The second seawater induced hydrogen evolution pathway is the following direct copper reaction with water [17]

$$2Cu + H_2O \rightarrow Cu_2O + 2H^+ + 2e^- \rightarrow Cu_2O + H_2 \quad (6)$$

Reaction (6) is suppressed in the absence of seawater because of the compact film of the cuprous oxide reaction product. Upon seawater exposure, chloride ions cause the formation of a cuprous chloride ion that partially dissolves the cuprous oxide film and forms the more porous Cu_2O and layer of atacamite ($Cu_2(OH)_3$- Cl), which readily admits more chlorine ions to reach the copper underneath it [17, 19]. The dissolution/porosity of the oxide film allows the oxidation reaction to continue and also results in further copper dissolution. Thus, copper corrosion can be often seen in the form of the green patina due to the cuprous chloride, and sometimes seen merely as a loss of copper without any discoloration. The latter was reported to be the case in the buried cable beach incident summarized above [16].

Reported test data on copper seawater corrosion rates can be used in conjunction with Reaction (6) to quantify the corresponding expected hydrogen generation rates. Corrosion rate data used for this purpose was obtained from copper exposure laboratory tests with natural and artificial and aerated seawater [19, 20], and from field tests with copper samples in seawater and sea aerosols in North Atlantic and tropical environments [21, 22]. The calculated hydrogen generation rates for exposures of several months varied from about 2 cc-H_2-per-hr-m^2-Cu to almost 14 cc-H_2 per hr-m^2. The cumulative distribution of the calculated results is shown in Figure 6. These results can be used to determine ventilation rates needed to prevent flammable hydrogen-air mixtures in electrical junction boxes or other small electrical enclosures in a seawater or sea atmosphere environment.

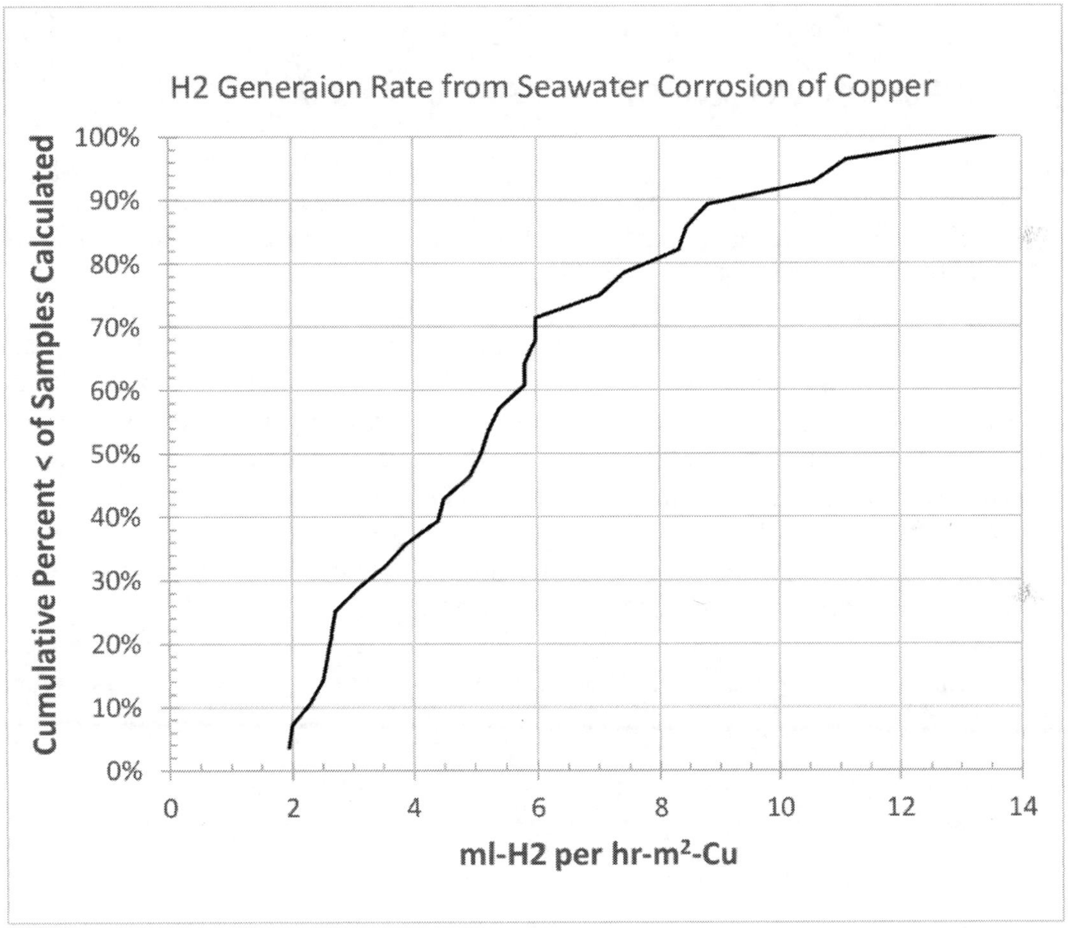

Figure 6. Calculated hydrogen generation rates from seawater exposure of copper.

4 Galvanized Sprinkler Piping Hydrogen Incidents

4.1 Helsinki Theater Sprinkler Room Explosion

On June 29, 2020, a sprinkler contractor was conducting routine maintenance on a dry pipe sprinkler system in the basement of a Helsinki theater. Shortly after the contractor opened the pipe, an explosion occurred that caused burn injuries to the contractor, who required hospitalization. The explosion caused minor damage to the stairwell adjacent to the sprinkler room [23].

An investigation by the Finnish Safety and Chemical Agency (Tukes) determined that the hydrogen was generated from a corrosive reaction involving zinc, water, and air within the galvanized steel pipe [24]. It was the first incident of its kind in Finland, but Tukes was sufficiently concerned about possible future incidents to issue a ban on the use of zinc-coated pipes in new sprinkler systems. Tukes also issued new requirements for maintenance involving opening of sprinkler piping; ventilation and a hydrogen detector are now required.

4.2 Other Sprinkler Pipe Hydrogen Fire and Explosion Incidents

Although the Helsinki theater explosion was the first incident in Finland involving hydrogen generated in galvanized sprinkler piping, there have been several such incidents in other Scandinavian countries, where galvanized dry pipe sprinkler systems are common. The following is a list of some European incidents.

- A 2014 incident occurred at a Danish industrial company when a sprinkler contractor was burned by a hydrogen flame vented from sprinkler piping while it was being drained [22].
- A second 2014 incident in Denmark; this one involved a hydrogen release from a pressurized water tank that had recently been drained. A sprinkler contractor suffered burn injuries from the resulting room explosion which may have been ignited by sparks from the contractors' tools [25].
- There was an explosion in a sprinkler fire pumphouse in France in October 2016 after a sprinkler contractor had drained the sprinkler piping in preparation for doing work on the system. The explosion occurred about 15 minutes after the drain was opened and the contractor left the pumphouse. Fortunately, there were no injuries, but the pumphouse was destroyed [26].

There have also been similar reported incidents in Japan. Yamate et al. [24] mention ignitions occurring when wet galvanized pipe have been cut or subjected to grinding during sprinkler system renovations.

4.3 Hydrogen generation reactions in sprinkler piping

The electrochemical reactions responsible for hydrogen generation in galvanized sprinkler pipe are analogous to the reactions described above for copper conductor exposure to seawater. The reactions described by Yamate et al. [27] are:

$$2H_2O \rightarrow 2H^+ + 2[OH]^- \quad (7)$$

$$2H^{+} + 2e^{-} \rightarrow 2H \rightarrow H_2 \quad (8)$$

$$Zn \rightarrow Zn^{2+} + 2e^{-} \quad (9)$$

$$Zn^{2+} + 2[OH]^{-} \rightarrow Zn(OH)_2 \quad (10)$$

Reactions (9) and (10) produce the zinc corrosion that results in formation of the white zinc hydroxide layer seen below the indicated water line in Figure 7. Su and Fuller [28] did not observe this corrosion when the air above the water was replaced by compressed nitrogen. This suggests that oxygen is required for accelerated corrosion in galvanized pipe, and that oxygen-water electrochemical reaction (5) is a key aspect of both corrosion and hydrogen production in galvanized dry pipe sprinkler systems in which there is residual water together with air in the piping. This often occurs when the piping is not completely drained following a dry pipe valve trip test or inadvertent actuation.

Figure 7 Zinc corrosion produced in 6 weeks in half-filled pipe [25]

Yamate et al. [24] have shown that rapid hydrogen generation can also occur from galvanic corrosion associated with water filled galvanized sprinkler pipe in contact with another metal component such as a stainless steel flexible pipe section. Gas samples from the galvanic coupling pipe revealed a hydrogen concentration of 20% and significant concentrations of zinc in the water. They also obtained visual evidence of hydrogen bubbles leading to a flammable atmosphere when the piping was opened, as shown in Figure 8.

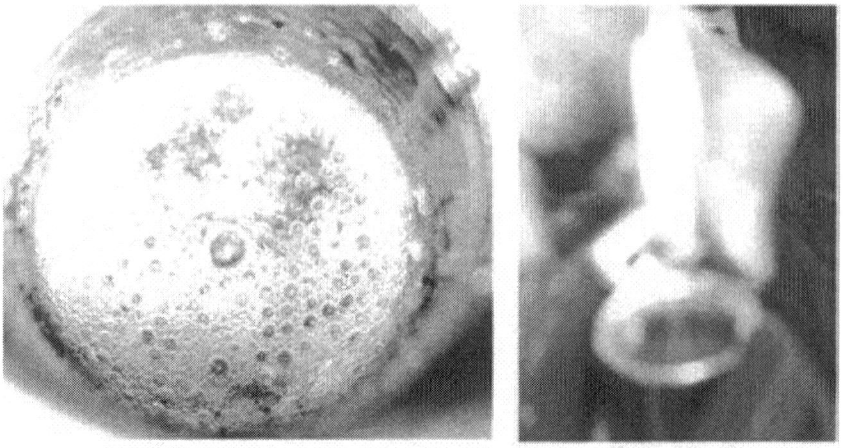

Figure 8 Hydrogen bubbles and flames in opened galvanic coupled piping [24]

Although the hydrogen generation incidents and accelerated corrosion observations have led to some bans on galvanized sprinkler piping, FM Global [28] notes that corrosion products in ordinary black carbon steel dry piping have been responsible for the vast majority of their fire losses due to obstructed sprinkler pipes. Therefore, FM Global continues to recommend use of galvanized steel pipe for dry and pre-action sprinkler systems. In view of the exposure test results noted above, FM Global recommends use of compressed nitrogen, rather than just air, in these dry pipe systems. This is different from the recommendations and requirements of many other organizations, as summarized in a recent Sprinkler Age article [29].

5 Conclusions

Metal-water chemical and electrochemical reactions have been responsible for hydrogen generation and subsequent ignition of hydrogen-air mixtures in many explosion incidents. Many incinerator ash-water incidents have occurred when the ash had a substantial aluminum and alkaline content. A compilation of test data has led to a statistical distribution of the expected hydrogen generation per unit mass of ash, with the median rate being greater for fly ash than for bottom ash.

Copper-seawater electrochemical reactions in offshore facility electrical junction boxes have led to hydrogen accumulations and subsequent ignitions associated with electrical connector arcs and sparks. Copper corrosion test data together with reported electrochemical reactions have allowed calculations of hydrogen generation rates per unit of copper cable surface area. Special cable features, such as hydrogen absorption additives, and flameproof junction box construction can help prevent and mitigate the consequences of these incidents.

Electrochemical corrosion reactions in galvanized dry pipe sprinkler systems containing residual water have led to many reported hydrogen fire and explosion incidents, particularly in Scandinavia, where there are many of these sprinkler systems. Testing has shown that the accelerated corrosion in galvanized dry pipe can be greatly reduced by replacing air in the pipe with compressed nitrogen, and by avoiding the galvanic corrosion resulting from contact between dissimilar metal components in sprinkler piping.

6 References.

[1] NFPA 484 Standard on Combustible Metals, National Fire Protection Association, 2019.

[2] Marine Accident Investigation Branch, Serious Marine Casualty Report Number 26/2017, December 2017.

[3] UN, Recommendations on the transport of dangerous goods, Manual of tests and criteria, United Nations, 5th Edition, 2009.

[4] https://h2tools.org/lessons/hydrogen-gas-explosion-municipal-refuse-incineration-facility. Accessed on 2/8/2021.

[5] http://www.shippai.org/fkd/en/cfen/CC1200025.html. Accessed 2/8/2021.

[6] S. Mizutani,, S. Sakai,, and H. Takatsuki, Investigation of hydrogen generation from municipal solid waste incineration fly ash, J Mater Cycles Waste Manag (2000) 2:16–23.

[7] R. Larsen, Energy recovery of metallic aluminium in MSWI bottom ash, Umea University M.S. Thesis, 2014.

[8] M. Arm, and J. Lindeberg, Gas generation in incinerator ash, 2015, Available at https://www.researchgate.net/publication/266467653. Accessed 2/8/2021.

[9] UK Highways Agency Interim Advice Notice 127/09, Appendix B: The Use of Foamed Concrete, October 2009.

[10] A. Nithiya, A.Saffarzadeh, and T. Shimaoka, Hydrogen gas generation from metal aluminum-water interaction in municipal solid waste incineration (MSWI) bottom ash, Waste Management 73 (2018) 342–350.

[11] Rupturing of 6.6kV Power Junction Box installed at an offshore oil production Well Head Tower, https://www.eng-tips.com/viewthread.dm?qid=434654. Accessed 2/12/2021.

[12] IEC 60079-7, Explosive atmospheres - Part 7: Equipment protection by increased safety "e", IEC, 2015.

[13] N. Douglas, Copper Loss from Electrical Conductors in a Subsea Environment as a Result of Low Insulation Resistance, Oilfield Technology, January 2019. Available at https://www.researchgate.net/publication/331951501.

[14] ITU-T Recommendation L.27, Method for estimating the concentration of hydrogen in optical fibre cables, International Telecommunication Union, 1997.

[15] State: Hydrogen from corrosion of copper cable caused Salty Brine beach blast, Providence Journal, July 24, 2015.

[16] Case Closed: Hydrogen Gas Combustion Caused Salty Brine Beach Blast, National Public Radio, July 24, 2015. Accessed 2/12/2021.

[17] E-S. Sherif, and S-M. Park, Inhibition of Copper Corrosion in 3.0% NaCl Solution by N-Phenyl-1,4-phenylenediamine, Journal of The Electrochemical Society, 152 10 B428-B433 (2005).

[18] P. Martins, et al, Hydrogen evolution reaction on copper: Promoting water dissociation by tuning the surface oxophilicity, Electrochemistry Communications 100 (2019) 30–33.

[19] E-S Sherif, R. Erasmus, and J. Comins, Corrosion of copper in aerated synthetic sea water solutions and its inhibition by 3-amino-1,2,4-triazole, Journal of Colloid and Interphase Science, 309 (2007) 470–477.

[20] F. Mansfield, et al., The Corrosion Behavior of Copper Alloys, Stainless Steels, and Titanium in Seawater, Corrosion Science, Vol. 36, No. 12, pp. 2063-2095, 1994.

[21] A. Drach, et al., Field studies of corrosion behaviour of copper alloys in natural seawater, Corrosion Science 76 (2013) 453–464.

[22] L. Nunez, et al, Corrosion of copper in seawater and its aerosols in a tropical island, Corrosion Science 47 (2005) 461–484.

[23] Explosion in Helsinki City Theater, https://www.foreigner.fi/content/print/explosion-in-helsinki-theater-at-least-one-person-in-hospital/20200629150657669I. Accessed 11/17/2020.

[24] Hydrogen explosion caused the Helsinki City Theatre incident, https://tukes.fi/en/-/hydrogen-explosion-caused-the-helsinki-city-theatre-incident. Accessed 2/14/2021.

[25] Explosions in sprinkler system probably caused by hydrogen gas, https://dpa-e.eu/explosions-in-sprinkler-system-probably-caused-by-hydrogen-gas/ Accessed 2/14/2021.

[26] Internally Galvanized Steel Pipe Blamed for Explosions, https://www.complianceonegroup.com/ lifesafety/ internally-galvanized-steel-pipe-blamed-for-explosions. Accessed 2/14/2021.

[27] T. Yamate, et al., Experimental Study on Hydrogen Generation Phenomenon in Fire Piping System in Buildings, Corrosion Engineering, 59, 436-439, 2010.

[28] P. Su, and D. Fuller, Corrosion and Corrosion Mitigation in Fire Protection Systems, FM Global Report, July 2014.

[29] Galvanized Steel Sprinkler Piping, Sprinkler Age, November 2018, Available at https:// sprinkle rage.com/ galvanized-steel-sprinkler-piping/ Accessed 2/15/2021.

AUTHOR INDEX